T0291881

CAMBRIDGE LIBRARY COLLECTION

Books of enduring scholarly value

Darwin

Two hundred years after his birth and 150 years after the publication of 'On the Origin of Species', Charles Darwin and his theories are still the focus of worldwide attention. This series offers not only works by Darwin, but also the writings of his mentors in Cambridge and elsewhere, and a survey of the impassioned scientific, philosophical and theological debates sparked by his 'dangerous idea'.

Charles Darwin as Geologist

During his famous Beagle voyage, Darwin collected rocks, fossils and other geological specimens. No previous geologist had amassed such a detailed set of data. He identified raised beaches and remains of marine organisms high above the sea, understanding their significance as evidence of the uprising of landmasses. He also witnessed an earthquake and volcanic eruptions, concluding that both are related to movements of molten rock deep in the Earth. In this 1909 lecture, Sir Archibald Geikie, then President of the Royal Society, outlines Darwin's geological findings and explains how these underpinned his developing ideas. We learn of Darwin's theory of coral reef formation, and his fascination with the activities of earthworms. Finally the lecture considers the importance of Darwin's geological studies in formulating his theory of evolution by natural selection, leading to his masterpiece On the Origin of Species.

Cambridge University Press has long been a pioneer in the reissuing of out-of-print titles from its own backlist, producing digital reprints of books that are still sought after by scholars and students but could not be reprinted economically using traditional technology. The Cambridge Library Collection extends this activity to a wider range of books which are still of importance to researchers and professionals, either for the source material they contain, or as landmarks in the history of their academic discipline.

Drawing from the world-renowned collections in the Cambridge University Library, and guided by the advice of experts in each subject area, Cambridge University Press is using state-of-the-art scanning machines in its own Printing House to capture the content of each book selected for inclusion. The files are processed to give a consistently clear, crisp image, and the books finished to the high quality standard for which the Press is recognised around the world. The latest print-on-demand technology ensures that the books will remain available indefinitely, and that orders for single or multiple copies can quickly be supplied.

The Cambridge Library Collection will bring back to life books of enduring scholarly value across a wide range of disciplines in the humanities and social sciences and in science and technology.

Charles Darwin
as Geologist

*The Rede Lecture, Given at the Darwin
Centennial Commemoration on 24 June 1909*

ARCHIBALD GEIKIE

CAMBRIDGE
UNIVERSITY PRESS

CAMBRIDGE UNIVERSITY PRESS

Cambridge New York Melbourne Madrid Cape Town Singapore São Paolo Delhi

Published in the United States of America by Cambridge University Press, New York

www.cambridge.org
Information on this title: www.cambridge.org/9781108002578

© in this compilation Cambridge University Press 2009

This edition first published 1909
This digitally printed version 2009

ISBN 978-1-108-00257-8

CHARLES DARWIN
AS GEOLOGIST

CAMBRIDGE UNIVERSITY PRESS
London: FETTER LANE, E.C.
C. F. CLAY, Manager

Edinburgh: 100, PRINCES STREET
Berlin: A. ASHER AND CO.
Leipzig: F. A. BROCKHAUS
New York: G. P. PUTNAM'S SONS
Bombay and Calcutta: MACMILLAN AND CO., Ltd.

CHARLES DARWIN
AS GEOLOGIST

THE REDE LECTURE

GIVEN AT THE DARWIN CENTENNIAL COMMEMORATION
ON 24 JUNE 1909

BY

Sir ARCHIBALD GEIKIE, K.C.B.
D.Sc., D.C.L. (Oxon.)

PRESIDENT OF THE ROYAL SOCIETY

Cambridge:
at the University Press
1909

Cambridge:

PRINTED BY JOHN CLAY, M.A
AT THE UNIVERSITY PRESS

CHARLES DARWIN AS GEOLOGIST

When the Vice-Chancellor honoured me with his invitation to give the Rede Lecture this year, he informed me that it would be included in the proceedings of the present Celebration. Although he left me free in the choice of a subject, it was obvious that the lecturer could hardly hesitate to select a theme which would have reference, more or less direct, to the illustrious Naturalist whose Centenary the University had resolved to commemorate. The nature and extent of Charles Darwin's contributions to biological science have been so often and so fully described, and his influence on almost all departments of human thought has been so amply recognised, that for the present, little more may seem to remain to be said on the subject until, in the course of time, a fresh review of his relations to the history of intellectual development may be called for. Nevertheless, I venture to think that there is one branch of his scientific labours, the value and significance of which have scarcely perhaps received the notice and appreciation to which they are entitled. It is apt to be forgotten that Darwin began his active scientific career as a geologist, that it was mainly to geological problems that the earlier years of his life were devoted, and that it was in no small measure from the side of geology that he was led into those evolutional studies which have given him so just a title to our admiration and gratitude, and have placed him so high among the immortals. I have therefore decided to ask your attention to-day to an outline of what he accomplished in geology, and of the relation of his studies in that science to the great problems of evolution with which his name is indissolubly associated.

In Darwin's younger days geology had hardly as yet completely vindicated for itself an acknowledged and assured place in the circle of the natural sciences. Those who then cultivated it could not agree among themselves upon its fundamental principles. They were divided into three distinct and antagonistic schools, between whom a long and sometimes acrimonious controversy was waged. On the one side, the Neptunists, or Champions of Water, maintained, as the prime article of their creed, that our planet was once enveloped in an universal ocean, from which the various rocks now to be seen in its crust were deposited as chemical and mechanical precipitates. They ridiculed the notion that the globe had a heated interior, and they regarded volcanoes as a late appearance in the earth's history, due to the spontaneous combustion of subterraneous beds of coal[1]*. On the other side the Vulcanists, or Plutonists, with Fire as their watchword, vehemently insisted on the important part which they believed had been taken by the globe's internal heat, whether in the form of

* The figures in the text refer to the Notes at the end.

volcanoes or by the subterranean intrusion of molten material into the solid crust[2]. But this school was split into two parties who opposed each other with hardly less vigour than they both showed towards the Neptunists. The one section proclaimed that the phenomena revealed by geology bear witness to a far greater intensity of action in former ages than now, and especially to the occurrence from time to time of gigantic cataclysms whereby the face of the planet was changed, whole tribes of plants and animals were destroyed, and place was made for the creation of new faunas and floras. Those who belonged to this division were known as the Catastrophic or Convulsionist school[3]. The other section received the name of Uniformitarian, inasmuch as they held that Nature's operations have always been carried on much as they are now, that instead of being marked by periodical frenzies of energy, her action has been generally uniform, and that thus the Present may be taken as the Key to the Past[4]. It is interesting to find that Darwin in his youth had personal relations with some of the leaders in each of these three camps.

When as a lad of sixteen he went to the University of Edinburgh he found himself at the headquarters of the Neptunists in this country. The Professor of Natural History there, Robert Jameson, had been trained at Freiberg under Werner, the great law-giver of the Neptunist cult, and for more than twenty years had been carrying on an active propagandism of Wernerian doctrines, which have long since been abandoned as illogical and absurd. Darwin has recorded that he found the lectures "incredibly dull." He particularly refers to an excursion to Salisbury Craigs, near Edinburgh, where, with volcanic rocks all around, he heard the professor actually declare that "a trap-dyke, with amygdaloidal margins and the strata indurated on each side, was a fissure filled with sediment from above, adding with a sneer that there were men who maintained that it had been injected from beneath in a molten

condition." The young student had sharp
enough eyes to see the true state of the case
and sufficient independence to follow his own
judgment in the matter. Well might he long
afterwards remark, "When I think of this
lecture I do not wonder that I determined
never to attend to Geology[5]."

But fortunately for the science this deter-
mination disappeared when a few years
afterwards he discovered that the Wernerism
which he had been taught at Edinburgh by
no means represented the real character and
aim of geological studies. He tells us that
during the last of the three years which he
spent at Cambridge he read Humboldt's
Personal Narrative and Herschel's *Prelimi-
nary Discourse on the Study of Natural
Philosophy,* and he adds that "no one or a
dozen other books influenced him nearly so
much as these two[6]." Herschel's admirable
little volume with its logical presentation of
scientific method, and Humboldt's glowing
pictures of tropical scenery and wide outlook
into all domains of science might well kindle

in him a keen desire to follow in the paths of
scientific travel.

It was in the spring of the year 1831, when
these influences were at work on Darwin's
mind, that he was persuaded by his sagacious
and accomplished tutor, John Stevens Henslow, "to begin the study of geology." His
first essays in field-work appear to have been
made in the summer of that year when, to
use his own words, he "worked like a tiger"
among the geological sections around Shrews-
bury, trying to fill in geological details upon a
map of that district—a task which proved less
easy than he expected[7]. Henslow introduced
him to Adam Sedgwick, the distinguished
Woodwardian Professor at Cambridge, with
whom he made a geological excursion that
autumn through a part of North Wales.
Doubtless practical lessons from so accomplished and inspiring a leader were of no
little service to Darwin in fostering his
growing geological enthusiasm and teaching
him the methods of observation in the field.
We may be sure, too, that he was privileged

to listen to much instructive and vigorous discourse, not only on the local geology, but on many of the wider bearings of the science. Sedgwick was a stout champion of the convulsionist creed, which he no doubt inculcated on his pupil as the true faith. That Darwin at first had himself some predilections in the same direction may perhaps be inferred from one of his letters to Henslow, in which, speaking of his geological doings in Shropshire before the excursion into North Wales, he said that he had "only indulged in hypotheses, but they are such powerful ones that, I suppose, if they were put into action but for one day, the world would come to an end[8]."

The first volume of Lyell's immortal *Principles of Geology* was published in the month of January 1830[9]. Although this work was eventually to have so profound an influence on the progress of geological science, it was received at first with doubt and opposition. In spite of its singularly luminous presentation of the whole connected system of natural operations that modify the surface

of the globe, alike in the organic and inorganic kingdoms, notwithstanding its overwhelming array of evidence drawn from all
quarters of the earth and the clear and
eloquent language in which the logical
deductions from this evidence were enforced,
the whole spirit and aim of the book ran so
directly counter to the tenets of the prevalent
convulsionist school that it seemed for a time
to reawaken the fires of controversy which
had for years been gradually waxing dim.
Its uniformitarianism, carried even to further
lengths than Hutton and Playfair had ven-
tured to go, was denounced with characteristic vigour by Sedgwick from the chair of the
Geological Society[4]. Even so judicious and
impartial a counsellor as Henslow, when he
advised Darwin to procure and study Lyell's
treatise, warned him "on no account to accept
the views therein advocated[10]." The young
geologist followed the advice of his tutor,
but not the proviso with which it was accompanied. He took the book with him on the
voyage of the *Beagle* and studied it with

enthusiasm, and with a result the very reverse of what Henslow desired. In recording, among the Cape de Verde Islands, the first observations made by him on foreign soil, he affirms that "they convinced me of the infinite superiority of Lyell's views over those advocated in any other work known to me [11]." In this way began that life-long indebtedness to Lyell which he sincerely felt, and never ceased to express.

In reviewing the nature and value of Darwin's geological work on the voyage of the *Beagle,* we must bear in mind the conditions under which it was undertaken. In the first place, he had only begun, a few months before, to turn his attention to the study of geology, and although he had doubtless worked hard at the subject during that brief interval, his knowledge and practical experience in it could hardly be other than limited. Further, it should be remembered that as the vessel was continually cruising from place to place, never remaining more than a short time at anchor, he had scant

opportunity of completing a detailed study
of any locality. His observations had often
to be made under the pressure of a limitation
in the number of days, or sometimes even
of hours, that were available. In the most
favourable circumstances, little more than a
first broad sketch of the geology of each
place could reasonably be expected. But
Charles Darwin was no ordinary observer.
From the very beginning of the voyage he
displayed a zeal and aptitude for geological
work which in a short time enabled him to
gather together an astonishing amount of
valuable detail, while at the same time he
rapidly gained experience in noting the wider
bearing of the facts that came under his eye,
and in drawing far-reaching and suggestive
conclusions from them.

The enthusiasm with which he pursued
his geological enquiries on the voyage is
charmingly revealed in his letters and his
Journal. The very first land at which the
Beagle touched on her outward voyage
gave him the opportunity of beginning his

field-work on the old volcanic ground of
St Jago, one of the Cape de Verde Islands.
Wandering over its mouldering lava-streams
and fired with fresh zeal from his eager
perusal of the first volume of the *Principles
of Geology* he realised, as he wrote in later
years, "the wonderful superiority of Lyell's
manner of treating geology, compared with
that of any other author whose works I had
with me or ever afterwards read[12]." It was
there under the spell of this great teacher
that, as he has recorded, "it first dawned on
me that I might perhaps write a book on the
geology of the various countries visited, and
this made me thrill with delight[13]." This
geological ardour lasted undimmed throughout the whole of the five years of the
Beagle's voyage. When after his first cross-
ing of the Atlantic he began his explora-
tions on South American soil, he wrote of
his voyage and its incidents: "Geology carries
the day; it is like the pleasure of gambling.
Speculating on arriving what the rocks may
be, I often mentally cry out '3 to 1 Tertiary

against Primitive'; but the latter have hitherto won all the bets[14]." In sending home his collections from Brazil he wrote to Henslow that "the box contains a good many geological specimens; I am well aware that the greater number are too small. But I maintain that no person has a right to accuse me till he has tried carrying rocks under a tropical sun. I have endeavoured to get specimens of every rock, and have written notes upon all[15]." After two years of continuous work along the American shores he could still tell the same sympathetic correspondent: "I am quite charmed with geology, but like the wise animal between two bundles of hay, I do not know which to like best—the old crystalline group of rocks or the softer and fossiliferous beds. When puzzling about stratification etc. I feel inclined to cry, 'a fig for your big oysters and your bigger megatheriums.' But when digging out some fine bones, I wonder how any man can tire his arms with hammering granite[16]."

How thoroughly he grew to be a field-geologist, never to be caught without his hammer, even when on botanical or zoological quests, is amusingly illustrated by his account of a visit to the island of San Pedro, north of the Chonos Archipelago off the southern part of the coast of Chile. Two of the officers of the ship had landed to take some bearings with the theodolite, while Darwin, hammer in hand, was rambling by himself. In the course of his perambulations, as he has told, "a fox (*Canis fulvipes*) of a kind said to be peculiar to the island and very rare in it, and which is a new species, was sitting on the rocks. He was so intently absorbed in watching the work of the officers that I was able by quietly walking up behind to knock him on the head with my geological hammer. This fox, more curious or more scientific, but less wise, than the generality of his brethren is now mounted in the Museum of the Zoological Society[17]."

While from time to time he was able to send home the collections that were accumu-

lating in the *Beagle*, his notes of specimens
and field-observations continued to grow in
bulk. By the middle of the voyage, as he
informed Henslow, they already filled some
600 quarto pages of manuscript, of which
about the half related to geology[18]. The
perusal of the second and third volumes of
the *Principles of Geology*, which had now
been published and were sent out to him,
increased his devotion to Lyellian views. He
even remarks that he was inclined to carry
some parts of the doctrine to a further length
than the master himself[19].

It is not possible within the limits of a
lecture to offer more than a mere sketch of
the geological work which was accomplished
by Darwin during the voyage. In four
distinct departments he enriched the science
with new and valuable material. In the first
place he added to our knowledge of the
volcanic history of the globe by many
detailed observations extending over a vast
geographical region. In the second place,
he brought forward a body of striking evidence

as to upward and downward movements
of the terrestrial crust, and drew from this
evidence some of the most impressive deductions to be found in the whole range of
geological literature. In the third place, he
made important observations on the geology
of South America, from the crystalline cores
of Brazil and the Andes to the Tertiary
and Post-tertiary deposits of Patagonia. In
the fourth place, he furnished new and interesting illustrations of the potent part taken
by the denuding agents of nature in effecting
the decay and degradation of the land.

I. As we follow Darwin in his rambles
over the volcanic tracts that were visited
during the voyage of the *Beagle*, we cannot
but be struck with the way in which he
always seeks to unravel the sequence of
events in the history of each centre of
eruption. While the details of rock-structure
and composition do not escape his notice,
their interest for him was obviously much
less than that of the chronicle of geological
changes which the rocks revealed. How

delighted he seems always to have been to
trace in a volcanic island the successive
phases from the early submarine eruptions
to the completed subaerial volcano! With
what zeal he observes and records the occurrence of layers of sea-shells and nests of salt
and gypsum, intercalated between the more
ancient lavas, as proofs of the uprise of the
sea-floor! The pursuance of the later history
of the volcano was a further source of keen
pleasure to him, as he noted the positions of the
earlier vents and the evidence of long intervals
of quiescence between the successive out-
bursts, when the mouldering lavas were
hollowed out by running water into valleys
and ravines, which eventually became the
channels wherein the youngest streams of
molten rock found their way to the sea. It
was always these broader questions of geological history that more especially appealed
to his imagination and awakened his active
interest.

Throughout those years of travel he was
ever on the outlook for fresh records of former

igneous action. As the result of his continued
observations he was able at last to sketch an
impressive picture of the part which that
form of geological agency has played in the
construction of the framework of the South
American continent. In his traverses of the
Andes, under an atmosphere resplendently
clear, and among steep mountain-sides rising
up absolutely bare of vegetation, he could
detect and follow the several great rock-groups
out of which the giant chain of the Cordillera
has been built. At the base he found a vast
succession of andesitic lavas and conglome-
rates, estimated by him to be some 7000 or
8000 feet thick, which from the fossils he
found associated with them, he assigned to
the age of the early part of the Cretaceous
period. Later in date came another copious
volcanic series in which the lower lavas were
believed by him to have been poured out
under the sea, while the later eruptions were
certainly subaerial, for he found a group of
erect silicified coniferous trees enclosed
among the hardened volcanic ashes and

surmounted by a massive canopy of basalt, 1000 feet in thickness[20]. Younger still were the huge basaltic sheets which he traced among the Upper Tertiary formations of the Santa Cruz river, and last of all came the existing still active volcanoes of the Andes, which brought the long record of eruptions down to the present day.

We may well believe that this marvellous chronicle of volcanic activity deeply impressed itself on Darwin's mind. It could not but give him a more vivid conviction of the potency of this branch of geological dynamics than most geologists have an opportunity of acquiring. Nevertheless it did not turn him into a convulsionist. Nor while his imagination dwelt upon the grand succession of events which revealed itself to him step by step as the years went on, did he neglect the less exciting but not less necessary observations of the lithological and other details that characterised the volcanic rocks. He was a diligent and judicious collector of rock-specimens, as his collections, still extant, abundantly testify, and he studied them with such appliances as were then available for petrographical research. Besides the elaborate notes which he made of their characters in the field, he examined them with the blowpipe, the goniometer, and the microscope, besides taking their specific gravity and applying to them the simpler chemical tests. His account of the bombs and the banded trachytes and obsidians of the island of Ascension has long taken its place as one of the classic descriptions of modern petrography[21]. Still more remarkable was his prescient inference as to the separation of the basic from the acid constituents in large bosses of granite—a suggestion which, after having been for many years lost sight of, has now been established as true[22]. His sagacious reflections upon the relations between the cleavage and foliation of altered rocks were likewise far in advance of his time[23].

II. The long series of observations and deductions made by Darwin on the move-

ments of the crust of the earth, both in an
upward and downward direction, have long
held an honoured place in the literature of
physical geology. He was the first observer
who could devote himself to this department
of investigation by personal research and
comparison over a vast area of the surface of
the globe, and could thus generalise in it upon
a basis of his own experience in the field.
During the very first halt of the *Beagle* at
the Cape de Verde Islands, his attention was
drawn to this subject of enquiry by the
evidence of upheaval which he met with at
St Jago. All through the successive years
of the voyage he continued to accumulate
facts, until they grew into such an array of
evidence as no previous geologist had ever
been able to amass. Especially important
were the proofs which he collected of the
rise of the southern part of South America.
From the shores of Brazil and Uruguay, where
the youngest marks of uplift are found only
a few feet above sea-level, he traced a magnificent succession of terraces that spread

over the broad tract of lowland between the
mountains and the Atlantic, and stretch
southwards for hundreds of miles to the
southern bounds of Patagonia. At least
eight of these terraces were noted by him,
each flanked with a bold line of winding
escarpment that fronted the coast-line and
slowly mounted as they were followed southwards, one above another, up to heights of
950 or even 1200 feet. The occurrence of
recent marine shells on at least the lower
platforms led him to the conclusion that the
uplift of this part of the continent must have
been a comparatively late geological event.
From the step-like series of terraces he inferred that the elevation took place inter-
mittently, with long pauses of rest, during
which the sea cut back the successive fronts
of these ancient inland-cliffs, as it is doing
still along the present coast-line. From the
greater height of the terraces in southern
Patagonia, he drew the deduction that the
uprise of the continent has been greatest
towards the south, and has gradually and

imperceptibly diminished in a northerly direction[24].

Similar evidence of the recent uprise of the continent was obtained by Darwin on the western coast at various places between lat. 46°35′ and 12°S.—a distance of more than 2000 geographical miles. The nearness of the mountain chain to the Pacific Ocean has not allowed the formation of any such display of broad platforms at successive levels, as in Patagonia. But he gathered conclusive proofs of uplift, not only in raised beaches with recent marine shells, but in abundant marks of old sea-margins at different levels, such as sea-worn caves, barnacle-crusted rocks, pholades still in their positions of growth, and successive gravel-terraces. The greatest height at which he was able to detect recent species of marine organisms was 1300 feet above the sea at Valparaiso[25].

As his acquaintance grew with the records of the geological history of South America, Darwin became more and more impressed by the proofs he obtained of the remarkable oscillations of level which the continent has undergone from the earliest times down to our own day. Reflecting on what he had seen on his traverses of the Cordillera and in Patagonia, he made this deliberate statement: "Daily it is forced home on the mind of the geologist that nothing, not even the wind that blows, is so unstable as the crust of the earth[26]."

It so happened that Darwin was ashore at Valdivia on the day of the great earthquake in February 1835, and felt the shock. A few days later the *Beagle* entered the harbour of Concepcion where, amidst a city of ruins, he came upon what he describes as "the most awful yet interesting spectacle he had ever beheld." He has recorded that from a geological point of view, "the most remarkable effect of this earthquake was the permanent elevation of the land," but he adds that instead of saying the effect, "it would probably be far more correct to speak of it as the cause[27]." He was satisfied that the land around the bay had been upraised two

or three feet, while on an island about thirty
miles off, Captain Fitzroy had found putrid
mussel-shells still adhering to the rocks ten
feet above high-water mark, where the inhabitants had previously dived at low-water
spring-tides for these shells. He connected
this result of the earthquake with the general
rise of the whole continent, regarding it as a
kind of sample of the process whereby the
uplift had been brought about. As he remarked in his *Journal*: "it is hardly possible
to doubt that this great elevation has been
effected by successive small upliftings, such
as that which accompanied or caused the
earthquake of this year, and likewise by an
insensibly slow rise, which is certainly in
progress on some parts of this coast[28]." As
his generalisation on the whole subject, he
held that "thousands of miles of both coasts
of South America have been upraised within
the recent period by a slow, long-continued,
intermittent, movement[29]."

This impressive conclusion, as the final
outcome of his long years of investigation,

was accepted by geologists and was incorporated by them into their common stock of ascertained knowledge. Some years ago, however, its validity was called in question. The illustrious president of the Vienna Academy of Sciences, Professor Suess, in the series of striking pictures which he has drawn of the changes which the surface of the earth has undergone, and of the causes to which these revolutions are to be ascribed, has referred to Darwin's observations, which he has somewhat summarily rejected as inadmissible. He has been led, I think, by his strong theoretical prepossessions against any kind of evidence for the secular elevation of continental areas of land, to minimise and explain away the proofs adduced by Darwin. He has availed himself of any expression of doubt or denial made by one or two later writers, which he accepts as well-founded. The testimony alleged to be borne by the terraces to the uprise of the land he briefly sets aside, with the suggestion that they may often be relics of the action of rivers or

lakes. The recent marine shells found inland
he looks upon as having been carried by
the inhabitants and to be counterparts of
the familiar kitchen-middens of European
coasts[30].

Charles Darwin was not a careless or casual
observer, nor one who rapidly jumped to a
conclusion from a limited basis of proof. He
was in the constant habit of repeating his
observations and checking his deductions,
and he had ample opportunities of doing so
in the geological field during the years that
he spent in South America. He was surely
competent to discriminate between platforms
extending for hundreds of miles parallel to
the coast-line, and terraces limited to each
river-system or to lakes. He was perfectly
familiar with the custom of the natives to
transport edible shell-fish for long distances
into the interior, and actually alludes to this
habit when describing deposits which he
believed to be true raised beaches[31]. He
was consequently on his guard against being
deceived by artificial accumulations of shells,

and he gives the criteria by which he dis-
criminated between them and natural deposits
—criteria which any field-geologist would
accept as sufficient.

Until therefore the evidence has been
sifted on the ground by a witness as capable
and as unbiassed as Darwin himself, I shall
continue to retain my belief in the trustworthi-
ness and importance of the observations and
conclusions of the great naturalist as to the
upheaval of those parts of South America
which he had himself the opportunity of examining. His contributions to this subject
have long been prized by geologists for their
fullness and clearness, and for their interest
and value in relation to the great problem
of the secular elevation of land. He himself
had no doubt that they were solid additions
to geological science, and such, I venture to
anticipate, will be the judgment of posterity.

After the close of the voyage of the
Beagle, when Darwin had found time to
study his collections and to reflect upon his
varied experiences of geological phenomena

during five busy years, he put in writing the matured opinions which he had formed on the forces concerned in continental elevation. His ample discussion of this subject, communicated to the Geological Society on March 7th, 1838, forms one of the most brilliant and suggestive essays which that Society ever published[32]. Although the progress of investigation has not sustained some important parts of his theoretical opinions on this subject, it is impossible to read his memoir without a high admiration for the genius of its author. Marshalling all the evidence then available, he arranges it in logical sequence and deduces from it conclusions of profound interest in regard to some of the obscurest problems in the history of our globe. It was the first attempt to treat this subject not as a mere matter of idle speculation, but on a basis of personal observation in the field. And thus, as a pioneering effort it is worthy of lasting recognition.

We can readily understand how he should

have been led to adopt the views promulgated in this remarkable paper. He had himself witnessed a severe earthquake, and could speak from personal knowledge of its effects in a region which had often been convulsed by similar events. He had found that one of these effects was a marked uplift of some parts of the coast-line. He had beheld with his own eyes the simultaneous and violent activity of two of the great volcanoes of the Cordillera[33]. Pondering on these mighty manifestations of terrestrial energy, and remembering what a long succession of volcanic periods he had detected in the framework of the continent, he conceived not only that earthquakes and volcanoes are intimately related to each other, as was then generally believed, but that they both proceed from movements in the internal molten material of the globe. Although the origin of these movements was shrouded from him, he became convinced, to use his own impressive words, that "the configuration of the fluid surface of the earth's nucleus is subject to some change—its cause completely unknown, its action slow, intermittent, but irresistible[34]."

These theoretical views seemed at the time to be warranted by all the evidence which had then been obtained on the subject, and more especially by the large body of proof which the author himself had gathered together. But the extended researches of later years in seismology and mountain-building have brought to light much information which he did not possess. We now know that there is no such general and intimate relation, as was then assumed, between earthquakes and volcanoes; for many gigantic earthquakes have taken their origin at a distance from active volcanoes, while vigorous volcanic energy is not always accompanied with earthquakes or with permanent alterations in the relative levels of sea and land. Since his time, too, the complicated structure of mountain-chains has been elucidated in much detail. We have learnt how intensely, along these tracts of elevated ground, the terrestrial crust has been folded, crumpled, fractured

and piled upon itself, without any sign of con-
comitant and co-operating volcanic agency.
But in regard to the cause of the secular uplift of continental land, we are still as ignorant
as Darwin confessed himself to be. It is
quite conceivable that for this phenomenon
his suggestion respecting movements of the
molten nucleus of the planet may, in some
form, come to be eventually established.

Besides meditating on the evidence in
favour of the elevation of land, Darwin during
his life in the *Beagle* had occasion to
consider terrestrial movements of an opposite
kind. It was during those eventful years
that he thought out his famous theory of
coral-reefs which gave to the world the most
original and impressive picture ever drawn
of the slow disappearance of an ancient land-
surface beneath the sea. The origin of these
singular islands, rising out of the profound
depths of mid-ocean, had long been a subject
of discussion, and several explanations of
them had been proposed, more or less plausi-
ble, but not free from objections. Darwin

offered a new suggestion which appeared to re-
move all the difficulties that were then known.
He showed how on the simple hypothesis of
a slow subsidence of the bed of the ocean,
fringing-reefs of coral along a coast-line could
be converted into barrier-reefs with a lagoon-
channel between them and the shore, and
further, how, where the land was insular and
continued to sink along with the surrounding
sea-floor, while at the same time the polypifers,
in their accumulation of calcareous material,
kept pace with the downward movement, the
barrier-reef would become an atoll or ring
of coral-rock enclosing a lagoon beneath which
the last peak of land might in the end disappear. With admirable clearness he worked
out the application of this theory to all the
facts that were then known about the structure and distribution of coral-reefs, and he
came to the conclusion that over vast spaces
of the Pacific and Indian Oceans former tracts
of land have slowly sunk beneath the water,
and that the sites of the submerged peaks

are to be recognised in the countless groups and archipelagoes of coral-islands[35].

The remarkable simplicity of this explanation of phenomena that had so long been matters of dispute, together with the grandeur of the vista which the theory opened up of a stupendous geographical revolution that had been in progress since a remote antiquity, assured Darwin's views of close attention and led to their general acceptance. First brought briefly before the Geological Society in 1837, and expounded more fully five years later in his well-known volume on coral-reefs, the theory held its place unchallenged for many years. Louis Agassiz had indeed insisted that it could not be applied to the coral-reefs of Florida, but not until 1863 were serious doubts thrown on its general applicability, when Professor Semper brought forward evidence of elevation among the Pelew Islands. In a second edition of his book, which appeared in 1874, Darwin briefly referred to this new evidence, but did not regard it as incompatible with his views. In

later years, however, the observations which
have multiplied over many widely distributed
parts of the Pacific and Indian Oceans, as
well as in the warmer waters of the Western
Atlantic, have supplied a large body of proof
that in many groups of coral-islands the
movement of the sea-bottom has been upward, the amount of elevation amounting in
some cases to more than 1000 feet. The
conclusion reached by such observers as Sir
John Murray, Professor Alexander Agassiz,
Dr H. B. Guppy and others is that true
atolls may be formed without subsidence,
by the outward growth of the coral upon a
talus of debris torn from the face of the reefs
by the force of the breakers. These writers,
who have carefully studied the subject on
the ground, have come to the conclusion that
Darwin's explanation cannot be maintained
as of universal application. After the fullest
consideration I have been compelled to
admit that this conclusion is well founded.
There can, I think, be no doubt that Darwin's
simple and striking explanation would per-

fectly account for the origin of a great many atolls. It remains to be seen whether, and how far, it may be possible eventually to discriminate between those which are to be thus understood from those where the coral-site has remained stationary or has been upraised. But the mere existence of an atoll can no longer be regarded as in itself a proof of subsidence. It has been to myself and to many other geologists a matter of keen regret that this brilliant generalisation of the great naturalist has been deprived of the wide application which for many years we attributed to it. But while we bow to the results of later investigation, we must still be allowed to regard it as a monument of his genius, which did good service by lifting geological speculation to a higher plane, and filling our minds with a more vivid conception of the gigantic scale on which the movements of the terrestrial crust may have been effected.

III. An important part of the solid work accomplished by Darwin during his life in the *Beagle* is to be found in his

numerous contributions to the elucidation
of South American geology. It would be
out of place to attempt to enumerate on
this occasion these various additions to our
knowledge of the subject. I have already
alluded to his studies of the older crystalline
rocks of that continent, to his sagacious conclusions regarding the connection between
cleavage and foliation which he drew from
these rocks, and to his far-sighted remarks
on the segregation of the more basic from
the acid constituents of eruptive bosses of
granite which he traced in Brazil. His traverses of the chain of the Andes enabled him
to furnish an interesting sketch of the general
architecture of that great range of mountains.
He fixed, from the evidence of associated
fossils, the geological age of the vast igneous
protrusions which form the core of the Cordillera. His researches in Patagonia led the
way in the investigation of the great Tertiary
series in that extensive territory. To his
enthusiastic labours we owe the important
palaeontological discoveries which for the first

time revealed the extraordinary abundance
and variety of the extinct vertebrate remains
in the youngest deposits of that region. Owen
in the Preface to his Memoir descriptive of
the series of fossils exhumed by Darwin,
speaks of the collection having been made
by one individual from a comparatively small
part of South America, and remarks that
"the future traveller may fairly hope for
similar success, if he bring to the search
the same zeal and tact which distinguish the
gentleman to whom Oryctological Science is
indebted for such novel and valuable ac-
cessions[36]."

IV. The voyage of the *Beagle,* with its
ample opportunities on land as well as on
sea, gave Darwin many occasions to study
the great system of agencies which are ceaselessly at work in sculpturing the face of
the land. He probably gained such a vivid
personal acquaintance with this subject as
few, if any, of the geologists of his day had
an opportunity of acquiring. This first-hand
knowledge stood him in good stead when in

later years he had to deal with questions of
geological time. It enabled him also to lend
a powerful support to the views of Lyell and
the cause of uniformitarianism against cata-
strophism.

The deep impression made on his mind
by the examples of stupendous denudation
which came before him in South America,
finds frequent mention in his writings. In
this regard, the chain of the Cordillera more
particularly roused his enthusiastic appreciation. "This grand range," he remarks,
"has suffered both the most violent disloca-
tions and slow, though grand, upward and
downward movements in mass. I know not
whether the spectacle of its immense valleys
with mountain masses of once-liquefied and
intrusive rocks, now bared and intersected,
or whether the view of those plains, composed
of shingle and sediment hence derived, which
stretch to the borders of the Atlantic Ocean,
is best adapted to excite our astonishment
at the amount of wear and tear which these
mountains have undergone[37]."

In the earlier part of his geological career,
like his great teacher Lyell, he was disposed
to credit the sea with a larger share than is
now generally believed to be its due in the
sculpture of the land. Nor need this be, in
his case, matter of surprise, for he had made
intimate personal acquaintance with the sea
alike in calm and in storm. He had seen
many striking instances of the efficacy of
breakers in the erosion of coast-cliffs. When
he visited St Helena and gazed on its range
of precipices rising here and there 1000 or
even 2000 feet above the waves that burst
into foam at their base, he felt that "the
swell of the Atlantic Ocean has obviously
been the active power in forming these cliffs."
Again as he sailed along the coast of Patagonia and traced its successive escarpments
that front the sea, one above another, for so
many hundreds of miles, he could not but be
impressed with the efficacy of marine action
in the denudation of that wide region. When
he found himself among the deep and wide
valleys of the Blue Mountains in New South

Wales, with their surrounding escarpment-cliffs, it was to the action of the sea that his thoughts naturally adverted as the cause of such a magnificent series of excavations. Like most of the geologists of the day he was convinced that "to attribute these hollows to alluvial action would be preposterous[39]."

Yet he was far from insensible to the results of the long-continued operation of sub-aerial agents in changing the face of the land. In his *Journal* he has recorded in graphic language the lesson on the erosive power of rivers which was graven on his memory by what he saw when he crossed the Andes by the Portillo Pass. As he watched the torrents, brown with mud, rushing headlong down the valleys and sweeping onwards the stones on their channels with a roar which could be heard at a distance, like the tumult of the sea in a storm, he realised how "the sound spoke eloquently to the geologist; the thousands and thousands of stones which, striking against each other, made the one dull uniform sound, were all hurrying in one

direction. It was like thinking on time, where
the minute that now glides past is irrecover-
able. So was it with these stones; the ocean
is their eternity, and each note of that wild
music told of one more step towards their
destiny. It is not possible," he continues,
"for the mind to comprehend except by a
slow process, any effect produced by a cause
repeated so often that the multiplier itself
conveys an idea not more definite than the
savage implies when he points to the hairs of
his head. As often as I have seen beds of
mud, sand and shingle, accumulated to the
thickness of many thousand feet, I have felt
inclined to exclaim that causes such as the
present rivers and the present beaches could
never have ground down and produced such
masses. But, on the other hand, when listen-
ing to the rattling noise of these torrents, and
calling to mind that whole races of animals
have passed away from the face of the earth,
and that during this whole period, night and
day, these stones have gone rattling onward
on their course, I have thought to myself can

any mountains, any continent withstand such waste[40]."

After an absence of almost five years the *Beagle* came back to England in October, 1836. That in spite of all the biological questions which during the voyage had shaped themselves before him and had engaged his keenest interest, Darwin still retained his early enthusiasm for geology is well shown in his records of the vessel's homeward journey which filled the fifth year of the expedition. It was during that year that he saw Tahiti, and touched at New Zealand, Sydney and Tasmania, everywhere adding fresh geological material to his notebooks. It was then, too, that he crossed the Indian Ocean and had an opportunity of making his study of coral-reefs which led to his generalisation about oceanic subsidence. On the same section of the voyage he again traversed the Atlantic twice, halting at St Helena and Ascension on the way, and once more landing at the Cape de Verde Islands as

the vessel finally shaped her course towards
home.   His letters show how eagerly, as each
chance presented itself, to use his own words,
he "set to work with a good will at my old
work of geology[41]."   From St Helena he
wrote to Henslow that he was "very anxious
to belong to the Geological Society[42]."   This
desire was speedily fulfilled.   His work on
the *Beagle* had become widely known by the
publication of excerpts from his letters to
Henslow.   His scientific reputation had consequently been so well established that not
only was he elected into the Society at the
beginning of the session in November, a few
weeks after his return, but in the following
February he was chosen as one of the Council,
and a year later (1838) was persuaded to
accept one of the two secretaryships—an
office which he held for three years.

He was now at the very centre of geological activity, surrounded with colleagues
whose names and work have given to that
heroic age of geology in this country an
imperishable lustre.   To be associated with

such leaders as Lyell, Sedgwick, Murchison, Greenough, Buckland, Fitton, De la Beche, Whewell and Owen could hardly fail to fan the flame of Darwin's geological proclivities. That he was appreciated and welcomed by these magnates in the science is testified in a hearty way by Lyell who wrote to Sedgwick (21st April, 1837): "It is rare even in one's own pursuits to meet with congenial souls; and Darwin is a glorious addition to my society of geologists, and is working hard and making way both in his book and in our discussions. I really never saw that bore —— so successfully silenced, or such a bucket of cold water so dexterously poured down his back as when Darwin answered some impertinent and irrelevant questions about South America[43]."

For some years most of Darwin's time was necessarily occupied in working up and publishing the voluminous material accumulated during his travels. Some of this material he prepared in the form of papers communicated to the Geological Society,

notably the great memoir, already alluded to,
on the Connection of Volcanoes and Earthquakes. But he found time also for some
fresh geological work in this country, more
particularly in regard to certain later phases
in the evolution of the present features
of the surface of the land. Thus in one of
these enquiries he was led to visit the Parallel
Roads of Glen Roy and to write a memoir
upon them wherein he advocated their marine
origin[44]. Somewhat later he made an excursion into the district of North Wales over
which Sedgwick had taken him eleven years
before. But in the interval the attention
of British geologists had been roused by
Agassiz to the proofs that their own country,
at a comparatively late geological period,
was buried under snow and ice. Darwin may
have been led to return to Caernarvonshire
by some vague recollection of topographical
features in that region which were not
specially noted by him at the time. He has
recorded that neither Sedgwick nor he "saw
a trace of the wonderful glacial phenomena

all around us. Yet these are so conspicuous
that a house burnt down by fire did not tell
its story more plainly than did this valley.
If it had been still filled by a glacier, the
phenomena would have been less distinct
than they now are[45]." The paper in which
he described his observations in this Welsh
valley was one of the earliest in the voluminous literature that has now gathered round
the subject of the glaciation of the British
Isles[46].

For some twenty years after his return
from the voyage of the *Beagle* Darwin continued to write occasional geological papers,
especially in relation to glacial matters, the
last of them being published so late as 1855[47].
Of all these contributions to geology the
most original and important was a brief
paper on the formation of vegetable soil,
which he communicated to the Geological
Society in the autumn of 1837[48]. The
youngest or surface layer of the earth's crust
had for many years been strangely neglected
by geologists. They had lost sight of the

pregnant reference to it made at the beginning of last century by Playfair. That far-seeing writer, following up the earlier ideas of Hutton, had pointed out how continually the surface soil is washed off the land and how it is as constantly renewed by the decay of organic and inorganic materials. But though he clearly recognised the reality and importance of this process of waste and renewal, he did not perceive the operation of perhaps the most important agency concerned in its efficiency. This discovery was first made known by Charles Darwin[49].

In the course of one of his visits to Maer Hall, his uncle, Josiah Wedgwood, called Darwin's attention to the curious way in which layers of cinders, burnt marl or lime, spread on the surface of pasture lands, eventually disappear under the grass, and at the same time suggested that this disappearance appeared to be due to the action of earth-worms in bringing up the finer particles of earth from below and leaving them on the surface. Darwin was naturally much interested in a subject so

obviously both geological and biological,
With his characteristic patience and care he
made a series of diggings, and soon satisfied
himself as to the facts to be accounted for.
He found that in one case a layer of marl,
spread over a field of pasture, had in about
80 years sunk some twelve or fourteen inches
beneath the surface. He came to the conclusion that this apparent subsidence had
undoubtedly been due to the continued
action of the worms, which after swallowing
and digesting the finer portions of the soil,
carry it up to the surface and void it there in
their castings. He drew the striking deduction that "every particle of earth forming the
bed from which the turf in old pasture-lands
springs has passed through the intestines of
worms."

Trifling as the topic may seem, and brief
as was the announcement of it (for the paper
filled only some four pages), the observations
published by Darwin were eventually seen to
possess a high importance in reference to
the problems of land-sculpture. But these

problems did not at that time, nor for many
years afterwards, engage much attention. It
was only when they began to be seriously
discussed, and when the evidence was accumu-
lating that the carving out of the face of the
land had not been in great measure the work
of the sea, as was so long believed, but was
mainly due to subaerial agencies, as Hutton
and Playfair had maintained, that the wide
significance of Darwin's little paper was perceived. It was then realised that even grass-
covered lands, screened as they seemed
effectually to be by their vegetable covering,
were nevertheless not exempt from the general
process of degradation, for it was manifest
that by the work of worms an appreciable
quantity of soil, brought up to the surface
every year, was there exposed to be washed
off by rain or to be dried and blown away by
wind. Thus level prairies and verdurous
slopes were seen to be no exception to the
operation of the universal ablation of the
land.

Although Darwin's original observations

on this curious and important subject remained, as it were, buried in the publications of a scientific Society, he never lost his interest in it. As he wrote to Professor Carus, "it had been to him a hobby-horse." He was accustomed to keep worms in pots, for the purpose of studying their habits, and eventually he was led to renew and extend the observations contained in his early paper. He attacked the problem in much greater detail than before, including, as part of his labour, minute investigations of the habits and mode of action of the worms. He likewise obtained more precise data by carefully measuring and weighing all the worm-castings thrown up within a given time in a measured space. The results of these patient enquiries were comprised in his well-known volume on Vegetable Mould[50].

It is interesting to remember that in this, his last published work, he returned once more to geological studies. But he now brought to their prosecution a wealth of biological experience and an ingeniously

4—2

devised system of measurement which gave
to his results a precision not always attain-
able in experimental geology. His volume
thus holds an altogether unique place among
modern contributions to the problems of
denudation. It shows no lessening in his
marvellous patience, his scrupulous aim at
accuracy and his masterly power of rising
from the minutest details into the broadest
generalisations. Geologists may well regard
this final volume as a legacy and example
to them.

I now come to consider in the last place
the geological side of Darwin's masterpiece—
The Origin of Species. This great work, the
outcome of his life-long researches and reflections, could not but contain frequent
reference to geological evidence which he
had himself gathered from so wide a field,
which he had pondered over so deeply, and
which was so intertwined with all his other
scientific work. We may compare his volume
to a great symphony in which the chords

from the various departments of biology are
blended into one vast harmony, but where
the deep under-tones of geology seldom fail
to be audible.

From the days of Buffon the problems
presented by the question of the geographical
distribution of plants and animals have
engaged the thoughts of many naturalists and
travellers. But not until the appearance of
Lyell's *Principles* were the geological aspects
of the subject systematically discussed. The
chapters in the second volume of that work,
wherein the phenomena of geographical distribution were shown to have so close a
connection with geological changes, must have
been diligently perused by Darwin on the
voyage of the *Beagle*. We may believe, indeed, that it was in no small measure from
their broad philosophical treatment and their
suggestiveness to him in his own researches,
that he conceived that deep respect and admiration for Lyell, to whom he was always proud
to acknowledge his indebtedness. Darwin's
two chapters on Geographical Distribution

bear the characteristic impress of that wide
biological and geological experience which
gave him so firm a mastery of the points to
be discussed. They display his candid fair-
ness in stating difficulties, together with his
earnest desire not to minimise or ignore them,
his caution and even diffidence in offering his
own suggestions for their solution, and his
power of luminous presentation wherewith he
could place the whole complicated subject in
coherent, intelligible and interesting form.
The progress of geology since Lyell's early
days enabled him to trace more definitely the
effects of geographical changes for which there
is reasonable evidence. Thus he attached
much importance to the direct and indirect
influence of the Glacial Period in reference
to the dispersal of plants and animals. His
treatment of this subject fills some of the
most striking pages of his volume. The reader
is made to realise, as he may never have done
before, that each species has had a long
geological history, which in many cases throws
light on the geographical revolutions that

preceded or accompanied the advent of man.

But with his cautious temperament he could find no favour for the bold hypotheses of some naturalists who, in default of other means of accounting for the present distribution of living organisms, have not scrupled to invoke the most gigantic changes in the disposition of sea and land, for which, however, no geological evidence can be adduced. "I do not believe," he affirmed, "that it will ever be proved that within the recent period most of our continents, which now stand quite separate, have been continuously, or almost continuously, united with each other and with the many existing oceanic islands[51]." He was content with less heroic methods of interpretation, and relied on such means of dispersal as can be seen to be effective in the present geographical condition of our globe.

The two specially geological chapters in the *Origin of Species* have always seemed to me to form Darwin's most momentous contribution to the philosophy of geology. I

well remember the effect which, when they
first appeared, they produced on at least the
younger geologists of the day. The fact that
the Geological Record is far from complete
was, of course, familiar knowledge. But until
these two chapters revealed it with such fullness of detail and such force of argument,
I do not believe that any one of us had the
remotest conception that the extent of its
imperfection was so infinitely greater than
we had ever imagined. The idea of progressive organic development was then in
general disfavour, and so long as that was
the case, the blanks in the Geological Record
lost much of their interest and importance
as indications of chronometric intervals.
Lyell, from the appearance of the first edition
of his *Principles of Geology* had consistently
maintained his determined opposition to all
doctrines involving the mutability of species.
In his ninth edition which, in an "entirely
revised" form, appeared in 1853, he could
still write: "the views which I proposed in
the first edition of this work, January 1830,

in opposition to the theory of progressive
development do not seem to me to require
material modification, notwithstanding the
large additions since made to our knowledge of fossil remains[52]." What he had been
inculcating for nearly a quarter of a century
had become the accepted belief of the great
body of geologists in this country, even of
those who dissented most strongly from his
uniformitarianism.

Yet there were some among them who
found it hard to follow their great leader in
this part of his teaching. He seemed to
them to undervalue the evidence that appeared so plainly to indicate that there has
been an ordered upward succession in the
appearance of the several divisions of the
animal and vegetable worlds. When he
declared that the occurrence of the remains
of fishes in the groups of strata below the
Coal formation "entirely destroys the theory
of the precedence of the simplest forms of
animals[53];" when he suggested that the non-
occurrence of mammalian remains among the

older rocks might be merely due to the
imperfect state of our information[54], and when,
in explanation of the poverty of the records
of the floras and faunas of the past, he offered
the consolation "that it has evidently been
no part of the plan of Nature to hand down
to us a complete or systematic record of the
former history of the animate world," an un-
easy conviction grew up that the testimony
of the rocks could not thus be set aside.
Vehemently insisting on "the doctrine of
absolute uniformity" in geological causation[55],
Lyell could account for the extinction of the
thousands of species of organisms that once
lived on the earth by reference to the normal
laws of nature, as seen in the operation of
the various causes that are still at work.
But when he contemplated the thousands of
new species which have successively replaced
those that died out or were destroyed, he
had recourse to a special act of creation for
each of them, thus appealing to an agency
whose working, while it might be in conson-
ance with natural law, lies outside of human

experience. His deliberate judgment was formulated in the following words: "Each species may have had its origin in a single pair, or individual where an individual was sufficient, and species may have been created in succession at such times and in such places as to enable them to multiply and endure for an appointed period, and occupy an appointed space on the globe[56]."

It is well to recall these aspects of geological thought in the middle of last century, and to remember what a dead weight of opinion, or, if we choose to call it prejudice, was opposed to the reception of Darwin's views. We must bear in mind also that the leader of this school of thought was none other than his own revered master Lyell, at whose feet he had sat for so many years and to whom he felt that he owed more inspiration than to any other man. Lyell, who had all his life opposed the idea of the mutation of species, was slow to be completely convinced of the truth of the conclusions at which his friend and follower had arrived.

In the volume on the Antiquity of Man which
he published four years after the appearance
of the *Origin of Species* he hesitatingly and
only partially accepted them[57]. In the course
of a few years more his conversion was complete, as he announced in the tenth and last
edition of his *Principles of Geology.*

Lyell's courageous abandonment of opinions which he had stoutly proclaimed through
a long life was a noble example of self-
abnegation in the cause of truth. It did
good service in helping forward the general
acceptance of the newer creed, and it was
especially appreciated by the younger geologists. Among their number were not a few
who felt when they read the *Origin of Species,*
that truly the scales had now fallen from
their eyes. There had been with them a
conviction that the grand progression of
organic life, from the earliest time until now,
must somehow have been governed by normal
biological law, though no satisfactory explanation had been offered of the manner
in which this continuous upward progress

had been achieved. Darwin's treatment of
the subject fascinated them by the genius
with which his long and varied experience
abroad and at home, alike in the geological
and biological domains, was brought to bear
on the elucidation of the great problem of
evolution for which he had so amply prepared
himself. Especially were they struck with his
mastery of the whole range of stratigraphy.
Into that department of geology he threw
a flood of new light, as for example when he
so cogently urged that complete conforma-
bility, or absence of visible break, may be no
proof of continuous deposition, but may conceal protracted periods of time unrepresented
by strata. Yet even where his arguments
were most forcible and convincing, they were
stated without the least show of dogmatism,
but with a quiet restraint that was apt to
conceal their strength. As we read them
now, they seem to be so obvious that it may
be wondered why they were not pressed long
before. Not only did he convince us of the
unsuspected degree of imperfection in the

Geological Record, but he revealed a new
method of interpreting it by showing that, on
the theory of descent with modification, fossils
possess a high chronometric value as in-
dicative of the relative importance of stratigraphical horizons and likewise a new suggestiveness in regard to geographical changes of
which no other memorial may have survived.

The light thrown by Darwin upon the
fossiliferous formations of the earth's crust
led to clearer conceptions of the principles
that must be applied to the interpretation
of the facts of stratigraphy. The sudden
appearance of whole groups of new species
upon a special stratigraphical platform, had
once been confidently appealed to as evidence
of a fresh creation of plants and animals to
replace those which were destroyed by a catastrophe that convulsed the world. This opinion,
though no longer expressed in the crude
shape in which such writers as Cuvier had
announced it[58], still in a modified form influenced many naturalists and geologists who,
though not convulsionists, were opposed to

the idea of the transmutation of species.
Darwin's cogent reasoning may be said to
have finally set it aside, by showing how such
breaks in the succession of organic remains
may be completely explained by regarding
them as marking enormous chronological
gaps in the records of what was nevertheless a continuous organic evolution. How
soon these fertile ideas in the *Origin of
Species* bore fruit was shown a few years
after the publication of that work, when
Ramsay gave his two brilliant addresses to
the Geological Society on breaks in the
succession of strata in Britain[59].

When the history of the progress of
science in the nineteenth century comes to be
written the views expressed in the geological
chapters of Darwin's great work, whether novel
or enforcing with new emphasis what had been
more or less clearly perceived before, will
be seen to mark a notable epoch in modern
geology. They have thoroughly permeated
the recent literature of the science, insomuch
that there is sometimes a risk that the student

who finds them so intimately incorporated
may lose sight of the source to which he
owes them. As one of the survivors of the
time when the *Origin of Species* appeared I
am glad to be privileged with this public
opportunity of acknowledging the deep debt
which the science of geology, in many of its
departments and in the whole spirit by which
it is now informed, owes to the life-long
labour of the author of that work. Geologists
are proud to claim him as one of themselves
and as one of the great masters by whom
their favourite science has been advanced.
In their name, therefore, I beg to offer at
this centennial celebration our tribute of
gratitude and admiration to the memory
of Charles Darwin.

NOTES

[1] The best account of the Neptunist doctrines is to be
found in Robert Jameson's *Treatise on Geognosy* (1808),
which forms the third volume of his *System of Mineralogy*.
Owing in large measure to the eloquence and personal
influence of Abraham Gottlob Werner, the great apostle of
this creed at the mining school of Freiberg in Saxony, these
doctrines (often known as Wernerian or Wernerism) enjoyed
a great vogue all over Europe in the later decades of the
eighteenth and the earlier of the nineteenth century. But
their prevalence rapidly diminished after his death in 1817,
especially when some of his more distinguished pupils, such
as L. von Buch and A. von Humboldt, abandoned them.
Jameson, however, who had studied under Werner,
remained longer unconvinced of the untenability of his
master's opinions. In 1804 he had become Professor of
Natural History at the University of Edinburgh. Four
years later (as if in rivalry to the Geological Society of
London, which was started in the previous year) he founded
the Wernerian Society, one main object of which was to
support and propagate the teaching of Freiberg. Even so
late as 1826, when Darwin attended his lectures, he was
still inculcating to his students the then discredited notions
of Werner as to the aqueous origin of igneous rocks. He

afterwards at a meeting of the Royal Society of Edinburgh
frankly acknowledged that he had been compelled to abandon
the distinctive tenets of Wernerism. It has not been
possible to recover the precise date of this recantation,
though both the late Sir Robert Christison and Professor
J. H. Balfour assured me that they had been present when
it was made. It must have taken place between the years
1826 and 1838, for in the latter year Hay Cunningham
published his Plutonist description of Salisbury Craigs,
which he said was " nearly that which the Professor now
delivers to his pupils " (*Essay on the Geology of the
Lothians*, 1838, p. 56, footnote).

    [2] It was in this country that the Vulcanist or Plutonist
creed was first clearly proclaimed by Hutton in his *Theory of
the Earth*, of which the first sketch was laid before the Royal
Society of Edinburgh in 1785, and the enlarged form in
two octavo volumes ten years later. The general principles
expressed in this work were after Hutton's death expounded
and enforced with admirable force and elegance by John
Playfair in his *Illustrations of the Huttonian Theory* (1802).
These two authors may be said to have laid the foundations
of the physical side of modern geology. While fully
Plutonist in their teaching they yet recognised, more vividly
than had ever been done before, the potent influence of the
aqueous and atmospheric influences which have ceaselessly
modified the surface of the globe.

    In this country the controversy between the two schools
of the Neptunists or Wernerians and the Plutonists
(Vulcanists) or Huttonians was prosecuted with much
vigour, while it lasted, but it had practically died out some
time before the middle of last century. One of the most
curious signs of its decay is to be found in the last volumes

of the *Memoirs of the Wernerian Society*, which are as frankly Plutonist as the previous volumes had been exclusively Neptunist. One cause of the cessation of the warfare is undoubtedly to be recognised in the ultimate influence of the Geological Society, which was founded in 1807 for the purpose of investigating the facts of geology rather than the advocacy of any theory regarding them. The continual advance of the doctrines taught by Hutton and Playfair is well indicated by the successive appearance of memoirs in scientific journals and also independent treatises which stand out as landmarks in the progress of geology, culminating in 1830 when the first volume of Lyell's *Principles of Geology* made its appearance. It is interesting to remember that in his relations to geological science Charles Darwin lived through this transition period. He had actually been a pupil of Jameson, the high priest of Wernerism in Britain, and he became one of the earliest and most effective followers of Lyell, the great prophet of Uniformitarianism.

[3] It was natural that the phenomena of geology, appealing powerfully to the imagination in their striking memorials of terrestrial revolutions, should favour the rise of the Catastrophist school. They seemed to require the operation of stupendous convulsions, and to be wholly inexplicable by the action of any forces now visible to human observation. It was only by degrees, and after Lyell's able advocacy, that the efficacy of apparently feeble causes, acting through long periods of time, came to be recognised. Among the champions of this school none was more eloquent and outspoken than Adam Sedgwick, the illustrious Woodwardian Professor at Cambridge. In the address which he gave to the Geological Society on quitting the presidential chair on

February 18, 1831, he vigorously criticised the uniformitarian doctrines which had been in the previous year
advocated with so much persuasive power by Lyell in the
first volume of his *Principles.* The following passages
may be quoted from this address.

"Though we have not found the certain traces of any
great diluvian catastrophe which we can affirm to be within
the human period ; we have, at least, shown that paroxysms
of internal energy, accompanied by the elevation of mountain-chains, and followed by mighty waves desolating whole
regions of the earth, were a part of the mechanism of nature."
"Volcanic action is necessarily paroxysmal ; yet Mr Lyell
will admit no greater paroxysms than we ourselves have
witnessed—no periods of feverish spasmodic energy, during
which the very framework of nature has been convulsed and
torn asunder. The utmost movements that he allows are a
slight quivering of her muscular integuments." *Proc. Geol.
Soc.,* Vol. I, pp. 307, 314.

⁴ The uniformitarian doctrines in geology were clearly
enunciated in Hutton's *Theory of the Earth* (see especially
Vol. II, pp. 205, 328, 467, 510, 547), and were admirably
expounded in Playfair's *Illustrations of the Huttonian
Theory,* wherein a large body of evidence was brought
forward in support of them. Twenty years later Karl E. A.
von Hoff began to publish his laborious chronicle of all the
geological changes recorded by man within the times of
human history (*Geschichte der durch Überlieferung nachge-
wiesenen natürlichen Veränderungen der Erdoberfläche,*
Vol. I, 1822 ; II, 1824 ; III, 1834 ; IV, 1840 ; V, 1841).
There could not have been gathered together a more impos-
ing array of proof of the nature and importance of the
vicissitudes of the earth's surface now in progress than is

contained within these meritorious volumes. But the cumulative effect of such changes, prolonged through vast periods of time, was not for some time realised by the general body of geologists. It was reserved for Lyell to point out the deductions that might logically be drawn from the large accumulation of evidence by his time available, and thus to place geology on a more solid foundation with a rightful claim to a higher rank than it had hitherto held among the observational sciences. Darwin, who had known something of its state in earlier days, never wavered in his conviction that Lyell had revolutionised the science of geology.

[5] *The Life and Letters of Charles Darwin, including an Autobiographical Chapter*, edited by his son, Francis Darwin. Three vols. London, 1887. Vol. I, pp. 41, 42. See also my *Founders of Geology*, 2nd edn., p. 329, for an account of another excursion to Salisbury Craigs, where the Plutonist notions were contemptuously rejected by one of the Wernerian faith.

[6] *Life and Letters*, Vol. I, p. 55.

[7] *Ibid.*, pp. 56, 189.

[8] *Ibid.*, p. 189.

[9] The title of this work is *Principles of Geology, being an Attempt to explain the Former Changes of the Earth's Surface, by reference to Causes now in Operation.* The first volume which, as stated in the text, appeared in January, 1830, was followed by the second in January, 1832, while the third and concluding volume was issued in May, 1833.

[10] *Life and Letters*, I, p. 72.

[11] *Ibid.*, I, p. 73.

[12] *Ibid.*, I, p. 62.

[13] *Ibid.*, I, p. 66.

[14] *Ibid.*, I, p. 233.

[15] *More Letters of Charles Darwin, a Record of his Work in a series of hitherto unpublished Letters,* edited by Francis Darwin (1903), Vol. I, p. 9.

[16] *Life and Letters,* I, p. 249.

[17] *Journal of Researches into the Natural History and Geology of the Countries visited during the voyage of H.M.S. "Beagle" round the World,* Chap. xiii., p. 280. The first edition of this work was published in 1839 as Vol. III of *The Narrative of the Surveying Voyages of Her Majesty's Ships "Adventure" and "Beagle" between the years* 1826 *and* 1836. The citations in these Notes are made from the second edition, published 1845 in Murray's *Colonial and Home Library.*

The Museum of the Zoological Society was dispersed many years ago, but the important parts of its contents were placed in the Natural History galleries of the British Museum. In answer to an application which I made to Mr S. F. Harmer, F.R.S., Keeper of Zoology in the Natural History Museum, Cromwell Road, for information as to Darwin's specimen, he has been so good as to send me the following particulars : "The fox to which you refer as having been killed by Charles Darwin with a geological hammer is in our collection. It is represented by a skin and a skull, and it is the type of *Canis fulvipes,* Waterhouse."

[18] *More Letters,* Vol. I, p. 14.

[19] *Life and Letters,* Vol. I, p. 263.

[20] *Geological Observations on South America, being the Third Part of the Geology of the Voyage of the "Beagle" during the years* 1832 *to* 1836. 1846. Pp. 175, 202, 232, 241, 247.

[21] *Geological Observations on the Volcanic Islands,*

*visited during the Voyage of H.M.S. "Beagle," together
with some brief notes on the Geology of Australia and the
Cape of Good Hope, being the Second Part of the Geology of
the Voyage of the "Beagle."* 1844. Chap. iii.

[22] One of the most striking conclusions contained in
Darwin's volume on "Volcanic Islands" is to be found in
his account of the great granitic masses of South America.
He not only perceived that there might be a "sinking of
crystals through a viscid substance like molten rock"
(p. 118) by virtue of their specific gravity being greater
than that of the base, but he inferred that in large plutonic
masses "a certain amount of separation of their constituent
parts has often taken place." He goes on to remark, "I
suspect this from having observed how frequently dykes of
greenstone and basalt intersect widely extended formations
of granite and the allied metamorphic rocks" (p. 123). He
had never examined a district in an extensive granitic region
without discovering such dykes. He thinks it probable
"that these dykes have been formed by fissures penetrating
into partially-cooled rocks of the granitic and metamorphic
series, and by their more fluid parts, consisting chiefly of
hornblende [or augite] oozing out, and being sucked into
such fissures." "We may admit, in the case of a great
body of plutonic rock being impelled by repeated movements
into the axis of a mountain-chain, that its more liquid
constituent parts might drain into deep and unseen abysses ;
afterwards, perhaps, to be brought to the surface under the
form, either of injected masses of greenstone and augitic
porphyry, or of basaltic eruptions. Much of the difficulty
which geologists have experienced when they have compared
the composition of volcanic with plutonic formations, will,
I think, be removed, if we may believe that most plutonic

masses have been, to a certain extent, drained of those comparatively weighty and easily liquefied elements, which compose the trappean and basaltic series of rocks " (p. 124). For two examples of the segregation of a basic periphery in granitic bosses see Messrs Dakyns and Teall, *Quart. Journ. Geol. Soc.*, xlviii (1892), p. 104; and Mr A. Harker, *op. cit.*, l (1894), p. 311 ; li (1895), p. 125.

[23] Darwin made a special study of this subject wherever he had an opportunity of examining slates and schists in South America, and his observations and conclusions have been confirmed by later observers. "I cannot doubt," he says, "that in most cases foliation and cleavage are parts of the same process: in cleavage there being only an incipient separation of the constituent minerals: in foliation, a much more complete separation and crystallisation" (*Geol. Obs. on South America*, p. 166). This is doubtless true in regard to such rocks as clay-slate, phyllite, &c., where the successive stages from uncleaved, through cleaved into foliated rocks and perfect schists can be traced. Darwin also conceived that "the planes of cleavage and foliation are intimately connected with the planes of different tension to which the area was long subjected before the cessation of the molecular movement" (p. 168). He saw that the direction of these planes was always parallel to the principal axes of elevation (p. 169).

[24] *Geological Observations on South America*, Chap. i.
[25] *Op. cit.*, p. 32.
[26] *Journal of Researches*, Chap. xv. p. 321.
[27] *Op. cit.*, Chap. xiv. p. 310.
[28] *Ibid.*
[29] *Geological Observations on South America*, p. 135.
He believed that "the excessively disturbed condition of

the strata in the Cordillera, so far from indicating single periods of extreme violence, presents insuperable difficulties, except on the admission that the masses of once liquefied rocks of the axes were repeatedly injected, with intervals sufficiently long for their successive cooling and consolidation " (p. 248).

[30] The passages in which Professor Suess controverts Darwin's views will be found in the first and second volumes of his *Antlitz der Erde* (Vol. I, pp. 95—105 and Vol. II, pp. 522—534 of the English translation of his work which has recently appeared under the title of *The Face of the Earth*, edited by Prof. Sollas). One particular observation has been especially singled out for criticism. Darwin has recorded that on the island of San Lorenzo, at a height of 85 feet above the sea, he found a bed two feet thick of recent marine shells, some of them with their insides incrusted with barnacles and serpulae. From this bed, amongst light corallines, horny ovule-cases of mollusca and roots of sea-weeds, he extracted some bones of birds, heads of Indian corn, a piece of woven rushes, and another piece of nearly decayed cotton string. He regarded these relics of human workmanship to have been contemporaneously embedded with the shells, and inferred that the land had here been upraised 85 feet since Indian man inhabited Peru (*Geol. Obs. on South America*, p. 49, and *Journal of Researches*, Chap. xvi. p. 370).

The same locality was visited a few years later by Prof. J. D. Dana, who published the following comment upon Darwin's deductions. "The argument [for an elevation of this coast] is urged with force and discrimination by Mr Darwin. My own observations have been confined to so small a part of the coast, that any opinion here expressed is entitled to but little weight, especially as I am unable to draw comparisons with the beds in other portions of the western coast alleged as similar in character. I may, however, frankly confess that the evidence does not seem to me to place the question beyond doubt." He then proceeds to state his "sources of doubt." These are (1) the occurrence of the shells in an irregular unstratified bed, just beneath or in the soil; (2) the absence of an inner cliff at the place. He thinks it more likely that the shells together with the relics of human occupation were accumulated by the Peruvians themselves, and he goes on to refer to the habits of the Patagonians and New Zealanders in transporting shell-fish from the sea-coast inland. He suggests that possibly a rush of waters over the land, such as is occasionally produced by an earthquake, might have been concerned in the spreading out of these remains, though without further examination, he does not feel ready to attribute the effects to this cause (*Report of United States Exploring Expedition* (1838—1842) *under C. Wilkes, Vol. X. Geology* (1849), p. 591).

Professor Suess assumes that what Darwin observed was merely one of the "kitchen middens" which occur at many localities along the margin of the coast. He remarks: "When Darwin visited these coasts in 1835 little was known as to the wide distribution of such remains. It must therefore have filled him with the greatest astonishment to meet with a thread, pieces of wicker-work and other traces of human activity in a deposit of sea-shells on the island of San Lorenzo, near Callao, at a height of 85 feet above the sea, and he may well, according to the state of knowledge at that time, have regarded it as a proof of recent elevation. Dana,

who visited the place some years later, has already explained the circumstance." He cites in a footnote the
passage from Dana's work above quoted. I leave any
impartial reader to judge whether the extremely guarded
statement of the American geologist " explains the circumstance" or justifies the summary rejection of Darwin's
observation.

When Darwin read the passages in Dana's volume,
which was published in 1849, he was naturally somewhat
indignant. Thus in writing to Lyell in December of that
year about the volume, he referred to Dana as "disputing
my conclusions without condescending to allude to my
reasons. Thus, regarding S. Lorenzo elevation, he is
pleased to speak of my ' characteristic accuracy,' and then
gives difficulties (as if his own) when they are stated by
me, and I believe explained by me" (*More Letters of
Charles Darwin*, Vol. II, p. 226).

In the passage above quoted Prof. Suess expresses
his opinion that at the time of Darwin's exploration in
South America little was known of the wide distribution
of sea-shells in the interior of the country by the inhabitants. But he has omitted to notice the references
to this mode of transport which are made by Darwin
himself, who positively says that he was "well aware from
what he had seen at Chiloë and in Tierra del Fuego, that
vast quantities of shells are carried during successive ages,
far inland, where the inhabitants chiefly subsist on these
productions" (*Geol. Obs. on South America*, p. 33).

One of the grounds on which Darwin convinced himself
that he was dealing not with kitchen-middens but with
natural deposits of marine origin, was derived from an
examination of the comminuted organic debris found filling

the shells and diffused through the enclosing earth. This material "was in considerable part composed of minute fragments of the spines, mouth-bones and shells of echini, and of minute fragments of chiefly very young *Patellae*, *Mytili* and other species" (*Ibid*). Not until he returned home and had an opportunity of unpacking and studying his collections, did he realise how fully this accumulation of comminuted organisms confirmed his conclusion as to "the marine origin of the earth in which many of the shells are packed. Considering these facts," he adds, "I do not feel a shadow of doubt that the shells, at the height of 1300 feet, have been upraised by natural causes into their present position" (*Ibid.* footnote).

The endeavour to account for all the shelly deposits of the coast of Chili as mere human refuse has been extended to the eastern side of the continent by various writers. The most recent author who has treated of this subject, Dr H. von Ihering, Director of the Museum of São Paulo, Brazil, in a detailed memoir entitled *Les Mollusques Fossiles du Tertiaire et du Crétacé Supérieur de l'Argentine*, which forms the whole of Vol. VII of the third series of the *Anales del Museo Nacional, Buenos Aires* (1907), refers to the abundant evidence of a former sea-margin, 30 to 40 metres above the present level of the sea, along the coast of Brazil and Argentina. He alludes to vast accumulations of shells in southern Brazil, called by the natives *sambaquis*, which sometimes form hills that rise 10 or 15 metres above the low marshy land, while in other places the equivalent deposits consist only of more or less clayey soil through which oysters and other shells are scattered. These *sambaquis* are sometimes formed entirely of valves of *Anamalocardia*, in other places exclusively of

oysters or of valves of *Corbula mactroides prisca.* Layers
of one or other of the shells may be seen to alternate in
some of the mounds. "Some authors," says the writer,
"compare these deposits to the kitchen-middens of Denmark,
but the conditions are quite distinct, since the shells are
never found mingled with bones of animals of the chase,
of fish, of wood charcoal, of fragments of pottery or other
human relics. The archaeological objects which are met
with in these deposits are only associated with the skeletons
which have there been buried" (p. 430). Dr von Ihering adds
that he need not enter into the details of this matter, as he
has already fully discussed it in several communications,
of which he subjoins a list. Dr Florentino Ameghino, so
well known for his numerous contributions to the geology
and vertebrate palaeontology of Argentina, has pointed out
the distinction between true kitchen-middens and natural
deposits of shells on the Patagonian coast (p. 432).

[31] *Geological Observations on South America,* p. 33.

[32] The title of this Memoir is "On the Connexion of
certain Volcanic phenomena and on the Formation of
Mountain-chains and the effects of Continental elevations"
(*Trans. Geol. Soc.*, 2nd Ser., v, 1840, pp. 601—632). In this
paper the author expresses the opinion that the earthquakes
of South America are " caused by the interjection of liquefied
rock between strata " (p. 615). He supposes that "the train
of connected volcanoes in Chili and the tract of coast upraised,
extending together for a length of more than 800 geographical
miles rest on a sheet of fluid matter " (p. 630). He believes
that "mountain-chains are only subsidiary and attendant
phenomena on continental elevations " (p. 623). He argues
that while "mountain-chains are the effects of continental
elevations, continental elevations and the eruptive force of

volcanoes are due to one great motive power now in progressive action ; therefore the formation of mountain-chains
is likewise in progress and at a rate which may be judged
of by either phenomenon, but most nearly by the growth of
volcanoes " (p. 629). He thinks this subterranean " power,
now in action, and which has been in action with the same
average intensity (volcanic eruptions being the index) since
the remotest periods, not only sufficient to produce, but
which almost inevitably must have produced, unequal
elevation on the lines of fracture " (pp. 624, 625).

[33] *Journal of Researches*, pp. 275, 291, 310.

[34] *Trans. Geol. Soc.*, 2nd Ser., v (1840), p. 631.

[35] Darwin first published his theory of coral reefs in a
brief statement read to the Geological Society in 1837
(Vol. ii of the Society's *Proceedings*, 1838, pp. 552—554),
with the title "On certain Areas of Elevation and Subsidence
in the Pacific and Indian oceans, as deduced from the study
of Coral-formations." The first edition of his book on the
subject appeared in 1842 as the First Part of the Geology
of the *Beagle*, with the title, *The Structure and Distribution
of Coral Reefs.*

The first note of objection to the general applicability of
Darwin's explanation appears to have been raised by
Professor Louis Agassiz in 1851 (*Bull. Mus. Comp. Zool.*,
Vol. i), who from his investigation of the Florida reefs came
to the conclusion that they furnished no evidence of
subsidence—an inference which was subsequently sup-
ported by the more detailed investigations of his son,
Prof. Alexander Agassiz, in a paper on the Tortugas and
Florida Reefs (*Trans. Amer. Acad.*, xi, 1883). More
important evidence in the same direction was published
by Prof. Carl Semper from the Pelew Islands in 1863

(*Zeitsch. Wissensch. Zoologie*, XIII, 1863, p. 558 ; *Verhandl. Physik.-med. Gesellsch. Würzburg*, 1868, and *Die Philippinen und ihre Bewohner*, 1869), and by Dr J. J. Rein from Bermuda (*Bericht. Senckenberg. Naturforsch. Gesellsch.*, 1869—70, p. 157). These writers insisted on evidence of uplift where, according to Darwin's view, there ought to have been depression.

In 1874 Darwin published a second edition of his volume on Coral-reefs, revised and in some parts almost re-written. It contained some additional matter, and took notice of Semper's criticism, but without attaching to it any great importance as necessitating a modification of the theory originally promulgated. He was well aware that in many places coral-reefs have been upheaved, and he cites examples of them. He contemplated the association of elevation with volcanic action, and the absence of active volcanoes over vast regions where coral-islands are numerous seemed to him a corroboration of his view that these areas are sinking.

But he was at this time in the full tide of the biological researches which engrossed his attention during the years that followed the publication of the *Origin of Species*, and it was hardly possible for him to keep in touch with the progress of geological enquiry. In the preface to the second edition of his *Coral Islands* he says that he might have greatly improved his map of the distribution of coral-reefs if he " had been better situated during the last thirty years, for hearing of recent discoveries in the Pacific, and for consulting charts published in other countries." He was probably unaware of the early objections of Louis Agassiz and of those made in later years by Dr Rein. It is doubtful also whether or not he became aware of the large body

of evidence which, after the publication of his second
edition, came from many widely separated localities with
constantly increasing force in opposition to his theory.
But even if this fresh information reached him, he was
content to let the matter rest where he had left it. A third
edition of his book was issued after his death under the
care of Professor Bonney in 1889.

In 1880, after the great voyage of the *Challenger* had
been carried out, Sir John Murray published a theoretical
explanation of the origin of coral-islands without the aid of
subsidence. Pointing out, as Darwin had already done (see
Note 51), that the oceanic islands are almost all of volcanic
origin and thus that no evidence from continental rocks can
be adduced in favour of the former existence of land now
submerged, he argued that the submarine ridges and peaks
which rise to various distances from the surface are probably due to the protrusion of volcanic materials. These
platforms, he conceived, might be brought in two ways
to the proper level at which reef-building polypifers could
live and grow. Those which rose above the sea-level could
be worn away by breakers and currents until they were
reduced to the lower limit of wave-action, while those
which lay at greater depths could be brought up to the
requisite level by the deposit upon them of the remains
of the calcareous pelagic organisms which swarm in the
upper waters of tropical seas. Thus, partly by erosion and
partly by the accumulation of organic debris, fitting building-
places could be furnished for the growth of corals. The
chief reef-builders flourish most vigorously on the outer
margin, amidst the play of the waves which are always
bringing them food. By the force of the breakers huge
blocks of the coral-rock are torn off the face of the reef.

These form a steep talus below, and on the top of this talus
the reef continues to grow outward (J. Murray, *Proc. Roy.
Soc. Edin.*, x, 1879—80, p. 505 ; xvii, 1889, p. 79. In the
Proceedings of the Royal Physical Society of Edinburgh,
Vol. viii, p. 1, I gave an account of the state of the question
up to the year 1884).

Strong support to these views has been given by
Professor Alexander Agassiz, who in his numerous and
extensive cruises has acquired a more extended and intimate knowledge of coral-reefs than any living naturalist.
His various published Reports afford an ample picture of
the structure and growth of these reefs all over the Pacific
Ocean as well as in the warmer waters of the western
Atlantic. The reader will find an index to the more
important contributions to the literature of coral-reefs on
p. 614 of the first volume of my *Text-book of Geology*. The
latest work of note is the voluminous Report on the borings
carried out on the Atoll of Funafuti, published by the
Royal Society (*The Atoll of Funafuti—Borings into a
Coral Reef and the Results; Being the Report of the Coral
Reef Committee of the Royal Society*, 1904. See also
A. Agassiz, *Mem. Mus. Comp. Zool.*, Harvard, Vol. xxviii
(1903), p. 212). The cores extracted from a bore sunk
on the reef of this atoll down to a depth of 1114 feet were
carefully studied by the most competent naturalists and
yielded reef-building genera from top to bottom. The base of
the calcareous mass was not reached, but its total thickness
was proved to be more than 1100 feet. If it could be
certainly shown that this mass consists of coral-rock in
its original position of growth this particular atoll would
demonstrate subsidence of the sea-floor, and could thus be
cited in support of Darwin's view. But if the mass is made

up of material broken by the waves from the face of the
reef during its slow seaward extension, or if it consists in
part of Tertiary limestone, it would not give any certain
proof of change of level.

If however we turn to the abundant and striking evidence of uprise among the coral islands of the Pacific and
Indian Oceans and the reefs in the western part of the
Atlantic, which has been brought to light in recent years
by A. Agassiz, H. B. Guppy and others, it is, I think,
impossible any longer to insist on the vast area of subsidence in these oceans which Darwin's theory required. He
undoubtedly pointed out a *vera causa* in subsidence, which
under the requisite conditions would give rise to the succession of different types of reef ending in true atolls. But
it must be admitted that the later explanation, while quite
compatible with the existence of local subsidence in different
areas, is in harmony with overwhelming evidence in favour
of elevation rather than depression among many oceanic
islands.

[36] *Zoology of the Voyage of H.M.S. Beagle*, Vol. I,
Fossil Mammalia described by Richard Owen, with a Geological Introduction by Charles Darwin, 1838. The hope
here expressed by the great comparative anatomist has
been abundantly fulfilled by the successful labours of later
investigators, especially those of the Argentine Republic
and of the expedition to Patagonia sent out from Princeton
University.

[37] *Geological Observations on South America*, p. 247.
See also pp. 136, 185—187.

[38] *Volcanic Islands*, p. 91.

[39] *Ibid.*, p. 136.

[40] *Journal of Researches*, Chap. xv., pp. 316, 317.

[41] *Life and Letters*, Vol. I, p. 265.

[42] *Ibid.*, Vol. I, p. 267.

[43] *Life and Letters of the Reverend Adam Sedgwick*, by John Willis Clark and Thomas McKenny Hughes, 1890, Vol. I, p. 484.

[44] "Observations on the Parallel Roads of Glen Roy and of other parts of Lochaber in Scotland, with an attempt to prove that they are of marine origin" (*Phil. Trans.*, 1839, pp. 39—82). In this paper Darwin had in his mind that the only conceivable barriers of the supposed lake or lakes must have consisted of rock or of detritus, and he rightly refused to believe the supposition that barriers of these materials could be admitted. The idea of barriers of ice had not then been suggested, and the only waters that seemed capable of accounting for the terraces were those of the sea. In the following year, however, Agassiz showed that Scotland must have been deeply buried in ice, and suggested that the Parallel Roads marked the levels of lakes that had been ponded back by glaciers (*Proc. Geol. Soc.*, III, p. 327 ; *Edin. New Phil. Journ.*, XXXIII, p. 217 ; *Atlantic Monthly* for June, 1864). When Mr Jameson's paper was published in which this view was completely demonstrated (*Quart. Journ. Geol. Soc.*, Vol. XIX, p. 235), Darwin frankly admitted his own explanation to have been erroneous (*More Letters*, Vol. II, pp. 188—193).

[45] *Life and Letters*, I, p. 58. The same simile was used in the *Origin of Species*, p. 330. The references in these Notes are to the sixth edition of the work.

[46] This paper bears the title "Notes on the effects produced by the Ancient Glaciers of Caernarvonshire, and on the Boulders transported by Floating Ice" (*Phil. Mag.*, Vol. XXI, 1842, p. 180). It is an interesting example of a

characteristic phase in the evolution of opinion regarding the phenomena of the Ice-Age. At first the so-called "Drift," also scattered boulders and striated rock-surfaces, were all attributed to powerful debacles produced by earthquake shocks whereby the sea was violently launched across the surface of the land. When the idea gained ground that ice had in some way helped in these operations, the superficial accumulations were still regarded as having been deposited in the sea, over which icebergs and floes transported materials from the land. Even when the presence of former glaciers among the mountains of Britain was admitted, the general distribution of ice-borne boulders over the face of the country was still attributed to the sea. In this paper of Darwin's (which followed a previous communication by Buckland "On the Diluvio-glacial Phenomena in Snowdonia and the adjacent parts of North Wales"), while the moraines with their boulders and the ice-worn domes of rock are recognised as manifestly due to valley-glaciers, the boulders lying scattered over the surrounding district and the till underneath them are spoken of as having been transported by floating ice when the mountains formed islets in the sea. In like manner, in his paper on the Parallel Roads he took for granted that the boulders in the Lochaber district had been distributed by floating ice. It was long before the efficacy of land-ice as an agent in the transport of erratics was adequately acknowledged.

⁴⁷ This was an article "On the power of Icebergs to make rectilinear, uniformly-directed Grooves across a submarine undulatory surface" (*Phil. Mag.*, x, 1855, p. 96). A growing disposition was then showing itself to doubt whether floating ice could mould itself upon an irregular rock-surface. The way in which on glaciated rocks the

striæ mount over the protuberances and descend into the
hollows of such surfaces was gradually coming to be recognised as the work of land-ice. Darwin still clung to the
older faith. He thought that icebergs can mould themselves more perfectly than glaciers on the rocks over which
they are driven, and "can slide straight onwards over
considerable inequalities, scratching and grooving the undulatory surface in long straight lines."

[48] "On the formation of Mould," *Proc. Geol. Soc.*, II
(1838), pp. 574—576 ; *Trans. Geol. Soc.*, 2nd Ser., v (1840),
pp. 505—510.

[49] Hutton was the first geologist who grasped the
general principle that although a layer of soil remains as a
covering on the land, its component particles are continually being washed off the surface while, in compensation,
fresh materials are added to it from the slow disintegration
of the rocks underneath (*Theory of the Earth*, Vol. I,
pp. 205, 210 ; II, pp. 93, 94, 95, 96, 184, 196, 202, 242,
244). The question was stated with characteristic clearness and precision by Playfair. "It is interesting to
observe," he remarks, "how skilfully nature has balanced
the action of all the minute causes of waste, and rendered
them conducive to the general good. Of this we have a
most remarkable instance in the provision made for pre-
serving the soil, or the coat of vegetable mould, spread out
over the surface of the earth." He points out that although
its materials are easily and continually washed away by the
rains and carried down by the rivers into the sea, it still
remains as a covering on the land, being augmented from
other causes. "This augmentation evidently can proceed
from nothing but the constant and slow disintegration of
the rocks. In the permanence, therefore, of a coat of

vegetable mould on the surface of the earth, we have a demonstrative proof of the continual destruction of the rocks ; and cannot but admire the skill with which the powers of the many chemical and mechanical agents, employed in this complicated work, are so adjusted as to make the supply and the waste of the soil exactly equal to one another" (*Illustrations of the Huttonian Theory*, § 103).

These conclusions, so vital for an intelligent comprehension of how a land-surface, even when covered with vegetation, does not wholly escape from degradation, were for many years ignored by later writers. It is true that Lyell, in commenting upon the passage above quoted from Playfair's treatise, supports its main contention, though he adds that it did not take into account the organic material supplied from the atmosphere (*Principles of Geology*, first edit., Vol. II, p. 188). How little importance was generally attached to the Huttonian view of this matter may be gauged from the language used by Sedgwick from the chair of the Geological Society, when he vehemently opposed the uniformitarianism maintained in the then recently published first volume of Lyell's *Principles*. "The destructive powers of nature," he said, "act only upon lines, while some of the grand principles of conservation act upon the whole surface of the land. By the processes of vegetable life, an incalculable mass of solid matter is absorbed, year after year, from the elastic and non-elastic fluids circulating round the earth, and is then thrown down upon its surface. In this single operation there is a vast counterpoise to all the agents of destruction" (*Proc. Geol. Soc.*, I, 1831, p. 303. This portion of Sedgwick's address was made the subject of some caustic remarks by Lyell in the second volume of his *Principles*, p. 197).

But the most absolute negation of the Huttonian
doctrine is to be found in the lectures given at the
Collège de France by Élie de Beaumont, the most distinguished French geologist of his day. He devoted a
special discourse to the subject, wherein he entered upon
a detailed endeavour to prove that neither the soil, nor the
general surface of the land, nor the beds of rivers, have
undergone any perceptible modification during the time of
human history. He came to the conclusion that while
there are many places on the surface of the globe where
degradation is continual and plainly visible, this waste is
appreciable precisely because elsewhere the vegetable soil
preserves its integrity almost intact during immense periods
of time. "The surface of the ground, covered with vegetation, remains without sensible alteration for thousands of
years. It is a fixed point, a zero from which the phenomena
can be measured, which advance with rapidity " (*Leçons de
Géologie Pratique*, tome I, 1845, p. 182). Many years ago
I pointed out the fallacy in this reasoning (*Trans. Geol.
Soc. Glasgow*, Vol. III, 1868, p. 170).

Élie de Beaumont made no reference to Hutton or
Playfair, nor to Darwin's paper, which had been published
three years before he lectured on the subject. That the
views of the English naturalist met with little favour among
geologists was shown when, in reviewing the progress of
geology, D'Archiac summarised with approval the observations and conclusions of Élie de Beaumont, but referred to
Darwin's views as a "*singulière théorie.*" "We fear," he
added, "that the learned English traveller has been too
much prepossessed by the importance of an organic influence,
which could have no effect save in low and damp meadows.
Cultivated lands, woods, high-lying meadows, afford no

support to this view.   The formation of the vegetable soil results from the simultaneous co-operation of mechanical and chemical agents, often helped by human industry" (*Histoire des Progrès de la Géologie*, tome I, 1847, p. 224). Darwin in his volume on Vegetable Mould took notice of both these French writers.   He remarks that D'Archiac "must have argued from inner consciousness and not from observation, for worms abound to an extraordinary degree in kitchen gardens, where the soil is continually worked" (p. 4).

[50] *The Formation of Vegetable Mould through the Action of Worms, with Observations on their Habits,* 1881.   The popularity of this volume was immediate.   We learn from *Life and Letters,* Vol. III, p. 218, "that in the three years following its publication, 8500 copies were sold,"—a sale relatively greater than that of the *Origin of Species.*

[51] *Origin of Species,* Chap. xii., pp. 324, 347.   He adduces his reasons for this belief, one of the most important being the geological argument that the almost universally volcanic composition of oceanic islands does not favour the admission "that they are the wrecks of sunken continents. If they had originally existed as continental mountain-ranges, some at least of the islands would have been formed, like other mountain summits, of granite, metamorphic schists, old fossiliferous and other rocks, instead of consisting of mere piles of volcanic matter."   His opinion was in favour of the view that the present continents and oceans have long remained in nearly the same relative positions (Chap. x., p. 288).

[52] *Principles of Geology,* ninth edit., 1853, p. 146. On the same page he affirms "No satisfactory proof has yet been discovered of the gradual passage of the earth

from a chaotic to a more habitable state, nor of any law of progressive development governing the extinction and renovation of species, and causing the fauna and flora to pass from an embryonic to a more perfect condition, from a simple to a more complex organisation." It may have been allowable to say that no "law of progressive development" had been discovered, but of the fact that a striking progressive advancement had taken place there could no longer be any doubt.

[53] *Op. cit.*, first edition, Chap. ix., p. 148.

[54] *Op. cit.*, ninth edition, Chap. ix., pp. 134 *et seq.*

[55] It is not always quite clear what "uniformity" implied in the creed of the uniformitarians. Lyell disclaimed that he " contended for the absolute uniformity throughout all time of the succession of sublunary events " (*Principles*, ninth edit., p. 149). He insisted that "the order of nature has from the earliest periods been uniform in the same sense in which we believe it to be uniform at present, and expect it to remain so in future " (*Ibid.*). But human experience embraced a mere fraction of geological time, and gave but a limited basis on which to determine what "the order of nature " is. It was this limitation which so roused the indignation of the Catastrophists. Lyell's own inclination evidently was against an admission that geological energy had ever been manifested on a more vigorous scale than has been witnessed by man. As he described himself, he was " a staunch advocate for absolute uniformity in the order of Nature " (*Life, Letters and Journals of Sir Charles Lyell*, Vol. I, p. 260).

[56] *Principles of Geology*, first edit., Vol. I, p. 124. This statement was repeated up to and including the ninth edition of the work.

[57] Lyell's *Geological Evidences of the Antiquity of Man,
with remarks on Theories of the Origin of Species by Vari-
ation* was published in 1863. It greatly disappointed
Darwin with its halting language, when from their inter-
course and discussions on the subject he had expected more
decided support (*Life and Letters*, Vol. III, pp. 8 *et seq*).
The tenth edition of the *Principles of Geology* appeared in
two volumes, the first in 1867 and the second in 1868. The
latter contained the author's full acceptance of Darwin's
views. As Mr Wallace truly remarked, "the history of
science hardly presents so striking an instance of youthful-
ness of mind in advanced life as is shown by this abandonment of opinions so long held and so powerfully advocated"
(*Quarterly Review*, April, 1869).

[58] This thesis was maintained by Cuvier in his *Theory
of the Earth*—a work which went through many editions,
and of which an English translation appeared under the
editorship of Robert Jameson. Cuvier's great contemporary,
Lamarck, on the other hand, disbelieved in the occasional
catastrophes and re-creations which the former so confidently
asserted. On the contrary, he looked on the succession of
life as having probably been unbroken from the beginning,
and he believed the existing faunas and floras of the globe
to be the lineal descendants and representatives of other
forms which have preceded them, and the remains of some
of which have been preserved among the stratified rocks of
the earth's crust.

[59] During his presidency of the Geological Society A. C.
Ramsay gave from the chair two addresses (in 1863 and
1864), wherein he discussed the character and meaning of
what he called "breaks in succession" among the stratified
formations of Britain, whether these interruptions are

marked by unconformabilities or by abrupt changes in fossil contents.   These brilliant addresses, printed in the *Quarterly Journal* of the Society, contained the first detailed and serious attempt to show the relative chronometric value of "breaks in succession," and gave strong support to the arguments maintained in the geological chapters of the *Origin of Species.*

CAMBRIDGE: PRINTED BY JOHN CLAY, M.A. AT THE UNIVERSITY PRESS.

Printed in the United States
By Bookmasters